The ESSEN

ELECTRONICS I

**Staff of Research and Education Association,
Dr. M. Fogiel, Director**

> This book covers the usual course outline of Electronics I. For more advanced topics, see "THE ESSENTIALS OF ELECTRONICS II".

**Research and Education Association
61 Ethel Road West
Piscataway, New Jersey 08854**

THE ESSENTIALS OF ELECTRONICS I ®

Printed in the United States of America

Library of Congress Catalog Card Number 87-61809

International Standard Book Number 0-87891-591-5

Revised Printing 1989

ESSENTIALS is a registered trademark of
Research and Education Association, Piscataway, New Jersey 08854

WHAT "THE ESSENTIALS" WILL DO FOR YOU

This book is a review and study guide. It is comprehensive and it is concise.

It helps in preparing for exams, in doing homework, and remains a handy reference source at all times.

It condenses the vast amount of detail characteristic of the subject matter and summarizes the **essentials** of the field.

It will thus save hours of study and preparation time.

The book provides quick access to the important facts, principles, theorems, concepts, and equations of the field.

Materials needed for exams, can be reviewed in summary form — eliminating the need to read and re-read many pages of textbook and class notes. The summaries will even tend to bring detail to mind that had been previously read or noted.

This "ESSENTIALS" book has been carefully prepared by educators and professionals and was subsequently reviewed by another group of editors to assure accuracy and maximum usefulness.

Dr. Max Fogiel
Program Director

CONTENTS

CONTENTS

3 THE BIPOLAR JUNCTION TRANSISTOR

4 POWER SUPPLIES

5 MULTITRANSISTER CIRCUITS

9 FREQUENCY RESPONSE OF AMPLIFIERS 81

CHAPTER 1

FUNDAMENTALS OF SEMICONDUCTOR DEVICES

1.1 CHARGED PARTICLES AND THE ENERGY GAP CONCEPT

The electron:

A) Negative charge = 1.60×10^{-19} coulomb

B) Mass = 9.11×10^{-31} kg

Hole – In a semiconductor, two electrons are shared by each pair of ionic neighbors through a covalent bond. When an electron is missing from this bond, it leaves a "hole" in the bond, creating a positive charge of 1.6×10^{-19} C.

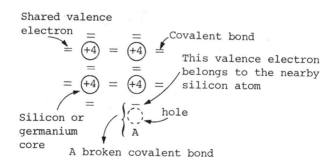

1

The energy gap concept and classification of materials

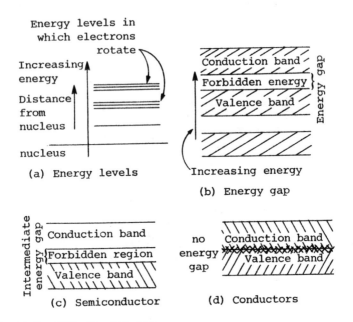

(a) Energy levels

(b) Energy gap

(c) Semiconductor

(d) Conductors

Drift and the Diffusion Current:

A) The diffusion current - The movement of charged particles due to a non-uniform concentration gradient.

B) The drift current - The movement of charges under the influence of an electric field.

1.2 FIELD INTENSITY, POTENTIAL AND ENERGY

Potential - The work done against an electric field in taking a unit of positive charge from point A to B.

$$v = - \int_{x_0 \text{ (point A)}}^{x \text{ (point B)}} E \cdot dx$$

2

Electric field intensity E:

$$E = \frac{-dv}{dx}$$

Potential energy U (in joules) equals the potential multiplied by the charge q.

Potential-energy barrier concept:

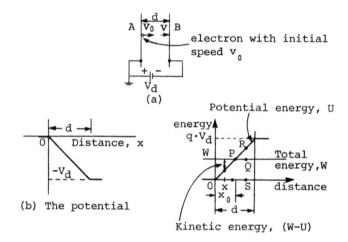

(a)

(b) The potential

electron with initial speed v_0

Potential energy, U

Total energy, W

distance

Kinetic energy, (W-U)

A) The kinetic energy is at its maximum when the electron leaves electrode A.

B) At P, no kinetic energy exists; the electron can therefore travel up to a distance x_0 from plate A.

The ev unit of energy: $1 \, ev = 1.60 \times 10^{-19} \, J$

1.3 MOBILITY AND CONDUCTIVITY

Mobility - When a metal is subjected to a constant E, a steady state is reached where the average value of the drift speed v is attained. v is proportional to E and is found by $v = \mu E$, where μ is the mobility of the electrons and

3

where the electric field has small values.

Current density: $J = \dfrac{I}{A} = \dfrac{N \cdot q \cdot \nu}{L\,A} = n \cdot q \cdot \nu = \rho \cdot \nu$

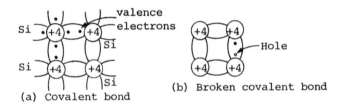

$\rho \equiv n \cdot q$ is the charge density; ν is the drift speed of the electrons.

Conductivity: $J = nq \cdot \nu = n \cdot q \cdot \mu \cdot E$

$J = \sigma E$, where $\sigma = nq \cdot \mu$ = conductivity of the metal

1.4 ELECTRONS AND HOLES IN AN INTRINSIC SEMICONDUCTOR

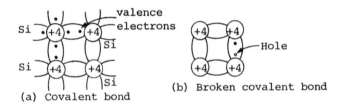

(a) Covalent bond

(b) Broken covalent bond

(c) Hole contributing to conductivity.

In an intrinsic semiconductor:

$n = p = n_i$ = density of the intrinsic carriers.

4

1.5 DONOR AND ACCEPTOR IMPURITIES

Donor impurity:

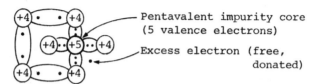

Pentavalent impurity core
(5 valence electrons)

Excess electron (free, donated)

Acceptor impurity:

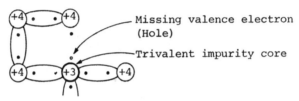

Missing valence electron
(Hole)

Trivalent impurity core

The mass-action law: $n \cdot p = n_i^2$

1.6 CHARGE DENSITIES IN A SEMICONDUCTOR

$N_D + p = N_A + n$, where N_D = positive charges contributed by the donor per meter^3, and N_A = negative charges contributed by the acceptor per meter^3.

$n \approx N_D$ (In an n-type material the free-electron concentration is equal to the density of the donor atoms.)

$p = \dfrac{n_i^2}{N_D}$ = Concentration of holes in the n type semiconductor

$p \approx N_A$ and $n = \dfrac{n_i^2}{N_A}$ (in a p-type material)

Generation and recombination of charges - On average, a hole or electron will exist for τ_p seconds or τ_n seconds, respectively, before recombination. This time interval is known as the "Mean Lifetime" of the hole or electron.

1.7 DIFFUSION AND THE POTENTIAL VARIATION WITHIN A GRADED SEMICONDUCTOR

Diffusion - The diffusion hole-current density J_p is, proportional to the concentration gradient.

$$J_p = -q \cdot D_p \cdot \frac{dp}{dx}$$

The Einstein relationship:

A) The Einstein equation: $\dfrac{D_p}{\mu_p} = \dfrac{D_n}{\mu_n} = v_T$

(B) $\boxed{V_T = \dfrac{\overline{K}T}{q}}$ = "Volt-equivalent of temperature"

$= \dfrac{T}{11,600}$ (T = temperature in °K)

C) At room temperature, $V_T = 0.0259V$ and $\mu = 38.6D$

Total current:

$$J_p = q \cdot \mu_p \cdot p \cdot E - q \cdot D_p \cdot \frac{dp}{dx}$$

$$J_n = q \cdot \mu_n \cdot n \cdot E + q \cdot D_n \cdot \frac{dn}{dx}$$

The potential variation in a graded semiconductor:

(A) $\boxed{E = \dfrac{V_T}{p} \cdot \dfrac{dp}{dx}}$ (E is the built-in field)

6

B) $E = \dfrac{-dv}{dx}$, hence $\boxed{dv = -V_T \cdot \dfrac{dp}{p}}$

C)

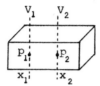

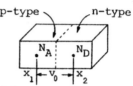

(a) A graded semi-
conductor, $p(x)$ is
not constant

(b) A step-graded jn

D) $V_{21} = V_2 - V_1 = V_T \cdot \ln \dfrac{p_1}{p_2}$

E) $\boxed{p_1 = p_2 \cdot e^{V_{21}/V_T}}$ This is the "Boltzmann relationship of kinetic gas theory".

F) Mass-action law: $n_1 = n_2 \cdot e^{-V_{21}/V_T}$ = Boltzmann equation for electrons

$$n_1 p_1 = n_2 \cdot p_2$$

G) An open-circuited, step-graded junction

a) $V_0 = V_{21} = V_T \cdot \ln \dfrac{p_{p_0}}{p_{n_0}}$, $p_1 = p_{p_0}$ = thermal-equilibrium hole concentration

in p-side and $p_2 = p_{n_0}$

= thermal-equilibrium hole in n-side

b) $p_{p_0} = N_A$ and $p_{n_0} = \dfrac{n_i^2}{N_D}$, such that

$$V_0 = V_T \cdot \ln \left[\dfrac{N_A \cdot N_D}{n_i^2} \right]$$

H) Analysis of p-n junction in thermal equilibrium.

Approximate doping profile for a p-n step junction:

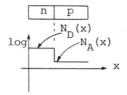

I) Unbiased p-n junction:

The equilibrium carrier concentration and potential as a function of distance in a p-n junction:

a) The region $x_n < x < x_p$ is called the "space-charge region" or "depletion region".

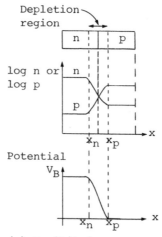

(a) Equilibrium
carrier construction
and potential

b) $$V_B = \text{built-in-voltage} = \frac{K \cdot T}{q} [\ln n_n - \ln n_p]$$

where

n_n = Electron concentration on n-side

n_p = Electron concentration on p-side

8

Depletion region width:

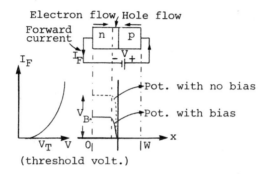

(a) Heavy doping (b) Light doping

J) Forward bias junction

The potential distribution in a p-n junction

n_w (electron concentration at $x = w$) $= n_{p0} = n_i^2/N_A$

n_{p_0} = Equilibrium electron concentration in p-region

$x = 0$ is the right-hand edge of the depletion region

n_A = Acceptor concentration

n_T = Electron concentration at $x = 0$

$$n_T = n_n \exp\left\{\frac{-q(V_B - V)}{KT}\right\}$$

$n_T = n_{p_0} \exp\left[\frac{q \cdot V}{KT}\right]$, n_{p_0} = electron concentration

Current density: $J_e = q \cdot D_e \cdot \frac{dn}{dx} = q \cdot D_e \frac{(n_w - n_T)}{w}$

$$= -q \cdot \frac{D_e n_i^2}{N_A w}\left(\exp\left[\frac{qv}{KT}\right] - 1\right)$$

9

The hole current is small compared to the electron current.

Reverse bias:

The potential distribution and electron concentration profile:

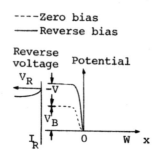

The total current flowing into the p-regin is:

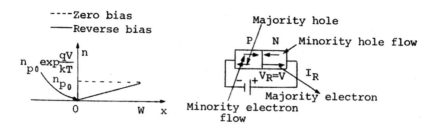

$$I = I_s (\exp \frac{q \cdot v}{KT} - 1) \quad \text{where } I_s = \frac{q \cdot D_e \cdot n_i^2 \cdot A}{w \cdot N_A}$$

CHAPTER 2

JUNCTION DIODE: THEORY AND SIMPLE CIRCUIT ANALYSIS

2.1 THE OPEN-CIRCUITED P-N JUNCTION

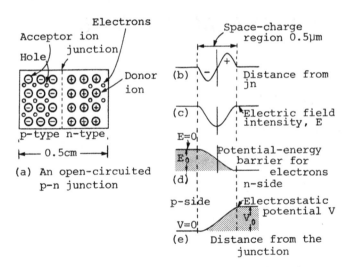

(a) An open-circuited p-n junction

(b) Distance from jn

(c) Electric field intensity, E

(d) Potential-energy barrier for electrons n-side

(e) Distance from the junction

Fig. 1 A schematic diagram of a p-n junction. Since potential energy = potential x charge, the curve in (d) is proportional to the energy for a hole & curve in (e) is proportional to the -ve of that in (d). (It is assumed that the diode dimensions are large compared with the space charge region.)

2.2 THE P-N JUNCTION AS A RECTIFIER

Reverse bias - The polarity is such that it causes both the holes in p-type and the electrons in n-type to move away from the junction. This cannot continue for long because the holes must be supplied across the junction from the n-type material.

A zero current would result, except that thermal energy present will create a small current, known as the "reverse saturation current," I_0. This current increases with increasing temperature.

Forward bias - In this case, the resultant current crossing the junction is the sum of the hole and the electron minority currents.

2.3 V-I CHARACTERISTICS

For a p-n junction, $I = I_0(e^{V/\eta \cdot V_T} - 1)$,

$V_T = \dfrac{T}{11600}$ = Volt-equivalent of temperature

The characteristics:

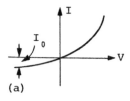

(a)

The reverse-bias voltage V_R is the voltage below which the current is very small (<1% of the maximum rated value).

2.3.1 TEMPERATURE DEPENDENCE OF THE V-I CHARACTERISTICS

$$I_0(T) = I_{01} \times 2^{(T-T_1)/10} \quad (I_0 = I_{01} \text{ at } T = T_1)$$

$$\boxed{\frac{dV}{dT} \approx -2.5 \text{ mv/}^\circ C}$$

2.4 DIODE RESISTANCE, TRANSITION AND DIFFUSION CAPACITANCE

Diode resistance:

The static resistance R is the ratio V/I.

For a small-signal operation, the "Dynamic-resistance" r is:

$$r = \frac{dV}{dI} \text{ (depending on the oeprating voltage)}$$

$$g = \frac{1}{r} = \frac{dI}{dV}$$

$$= I_0 \frac{\exp\left[\frac{V}{nV_T}\right]}{nV_T}$$

$$= (I + I_0)/n \cdot V_T$$

For $I \gg I_0$,

$$\boxed{\gamma \approx \frac{n \cdot V_T}{I}}$$

Piecewise linear characterization of a semiconductor diode

13

The break-point is not at the origin but at a point V_y units from the origin. V_y is called the offset voltage.

R_f is the forward resistance.

Transition and diffusion capacitance:

In the reverse bias region, the transition capacitance predominates, while in the forward bias region, the diffusion or storage capacitance predominates.

Transition capacitance, C_T:

$$C_T = \left| \frac{dQ}{dV} \right|$$

Note: C_T depends on the magnitude of the reverse voltage, since the magnitude of the reverse voltage determines the depletion width.

Diffusion capacitance, C_D:

C_D is introduced when the p-n junction is forward biased due to the additional injected charge redistribution in the n-region.

This type of capacitance limits switching speed in logic circuits used as junction devices.

Static derivation of C_D:

$$C_D = \frac{dQ}{dV} = \tau \cdot \frac{dI}{dV} = \tau \cdot g = \frac{\tau}{r}$$

$$C_D = \frac{\tau \cdot I}{\eta \cdot V_T}$$

For reverse bias, g is very small and $C_D \ll C_T$.

For a forward current, $C_D \gg C_T$.

$$r \cdot C_D = \tau$$

The diode time constant equals the mean lifetime of minority carriers.

2.5 ANALYSIS OF SIMPLE DIODE CIRCUITS: CONCEPT OF DYNAMIC RESISTANCE AND A.C. LOAD LINE

2.5.1 THE D.C. LOAD LINE

The behavior of the diode at low frequency is marked by V-I characteristics.

Other elements of a circuit beyond the region bounded by the diode and its terminals can be replaced by a thevenin equivalent circuit.

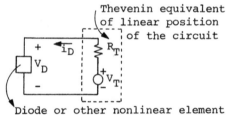

Thevenin equivalent of linear position of the circuit

Diode or other nonlinear element

For the non-linear element $i_D = f(v_0)$, the thevenin equivalent is given as $v_D = v_T - i_D \cdot R_T$.

The problem is solved by plotting these equations on the same set of axes.

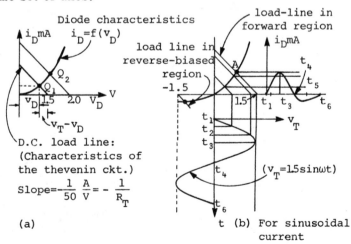

Diode characteristics

load-line in forward region

load line in reverse-biased region

D.C. load line: (Characteristics of the thevenin ckt.)

$\text{Slope} = -\frac{1}{50} \frac{A}{V} = -\frac{1}{R_T}$

(a)

$(v_T = 1.5 \sin \omega t)$

t (b) For sinusoidal current

The straight-line characteristics of the thevenin circuit is "D.C. load line." As long as R_T remains constant, any change in v_T is accounted for by a horizontal shift of the load line.

If v_T is sinusoidal, the corresponding current i_D can be found as shown in (b).

Small-signal analysis, dynamic resistance:

Small-signal - When the total peak-to-peak swing of the signal is a small fraction of its D.C. component.

$$v_T = V_{dc} + v_i = V_{dc} + V_{im} \sin \omega t$$

(where V_{dc} = bias voltage, and $V_{im} << V_{Dc}$).

The operating point for $v_T = V_{dc}$ is called "the quiescent point". It is found as follows:

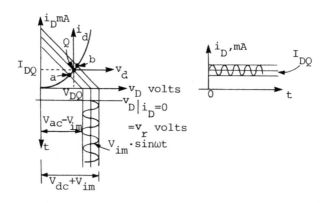

A new set of axes $i_d - v_D$ is constructed at Q (as shown below);

$$i_d = i_D - I_{DQ}, \quad v_d = v_D - V_{DQ}$$

16

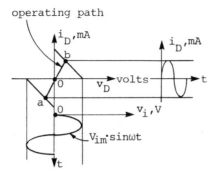

operating path

The operating path is "ab". The dynamic resistance r_d of the diode is equal to the inverse of the slope of line "ab".

$$r_d = \text{dynamic resistance} = \left. \frac{\Delta v_D}{i_D} \right|_{Q \text{ pt.}}$$

If r_d is found, circuit variables are obtained by using Ohm's law.

The original circuit has two parts:

(a) For calculating Q   (b) For calculating
                          small-signal
                          a.c. component

Calculation of r_d:

$$r_d = \left. \frac{d v_d}{d i_D} \right|_{Q \text{ point}} \cong \frac{V_T}{I_{DQ}} = \frac{2SmV}{I_{DQ}} \quad (\text{at } T = 300°K)$$

Reactive elements:

$$I_{dm} = (\text{the peak current}) = V_{im} / |r_i + r_d + z_i|$$

17

2.5.2 THE A.C. LOAD LINE

For the circuit shown below, the D.C. load line and Q point are:

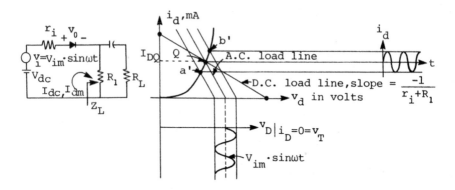

When the A.C. signal is present, the effective resistance seen by the diode is $r_i + (R_1 \| R_L)$. The A.C. load line is drawn through Q with slope $= \dfrac{-1}{[r_i + (R_1 \| R_L)]}$

The equations for the D.C. and A.C. load lines are:

$$V_{dc} = I_{DQ}(r_i + R_1) + V_{DQ} \qquad \text{D.C. load line}$$

$$v_i = i_d(r_i + [R_1 \| R_L]) + r_d \qquad \text{A.C. load line}$$

2.6 SCHOTTKY AND ZENER DIODES

Schottky diode:

It is formed by bonding a metal (platinum) to n-type silicon. A Schottky diode has negligible charge storage and is often used in high-speed switching applications.

Platinum acts as an acceptor material for electrons when bonded to n-type silicon.

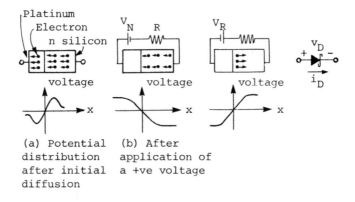

(a) Potential distribution after initial diffusion

(b) After application of a +ve voltage

Zener diode:

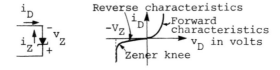

Reverse characteristics

Forward characteristics

v_D in volts

Zener knee

Unlike regular diodes, when you apply a high, reverse voltage across a Zener diode, you produce an almost constant-voltage region on the characteristic curve of the diode. One application of this special property of the Zener diode is voltage regulation.

The change of the Zener voltage V_Z as a result of a change of temperature is proportional to the Zener voltage as well as change in temperature.

$$T_c = \text{Temperature coefficient} = \frac{\Delta V_Z / V_Z}{\Delta T} \times 100\%/^\circ C$$

2.7 DIODE LOGIC CIRCUITS

2.7.1 A DIODE "AND" GATE

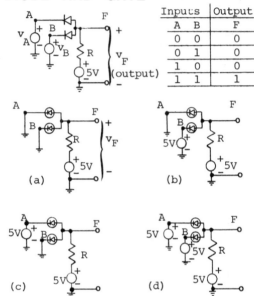

Inputs		Output
A	B	F
0	0	0
0	1	0
1	0	0
1	1	1

(a)   (b)

(c)   (d)

The diode AND gate redrawn for each possible combination of inputs. The diodes are assumed to be perfect rectifiers.

2.7.2 A DIODE "OR" GATE

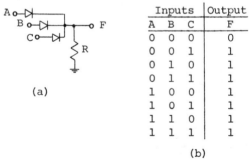

Inputs			Output
A	B	C	F
0	0	0	0
0	0	1	1
0	1	0	1
0	1	1	1
1	0	0	1
1	0	1	1
1	1	0	1
1	1	1	1

(b)

A diode OR gate. (a) The circuit. (b) The truth table. The output is 1 if any of the inputs is 1.

20

CHAPTER 3

THE BIPOLAR JUNCTION TRANSISTOR

3.1 THE JUNCTION TRANSISTOR THEORY

3.1.1 OPEN-CIRCUITED TRANSISTOR

Under this condition, the minority concentration is constant within each section and is equal to its thermal-equilibrium value n_{p_0} in p region and p_{n_0} in n region. The potential barriers at the junctions adjust to the contact difference of potential V_0, such that no free carriers cross a junction.

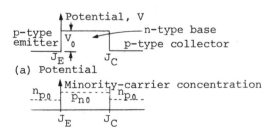

(a) Potential

3.1.2 TRANSISTOR BIASED IN THE ACTIVE REGION

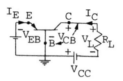

(a) p-n-p
transistor
biased in the
active region.

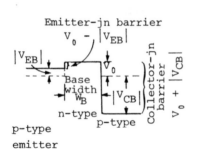

(b) Potential variation
through the transistor.

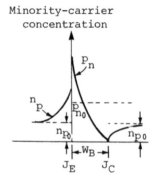

Minority-carrier
concentration

(c) Minority-carrier
concentration

The dashed curve applies to the case before the application of external biasing voltages.

The forward biasing of the emitter junction lowers the emitter-base potential barrier by $|V_{EB}|$, permitting minority-carrier injection. Holes are thus injected into the base and electrons into the emitter region.

Excess holes diffuse across the n-type base, the holes which reach J_c fall down the potential barrier and are collected at the collector.

3.1.3 TRANSISTOR CURRENT COMPONENTS

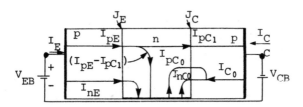

Transistor current compoments for a forward-biased emitter junction and a reverse-biased collector junction.

I_{pE} - Current due to holes crossing from emitter into base is the forward injection.

I_{nE} - Electrons crossing from base into the emitter.

$I_E = I_{pE} + I_{nE}$ (I_{pE} has a magnitude proportional to the slope at J_E of the p_n curve. Similarly, I_{nE} has a magnitude proportional to the slope at J_E of the n_p curve.)

$-I_{co} = I_{nco} + I_{pco}$ (I_{nco} consists of electrons moving from the p to the n region across J_c, and I_{pco} results from holes across J_c from n to p.)

I_c = Complete collector current = $I_{co} - I_{pc}$

$$\boxed{I_c = I_{co} - \alpha \cdot I_E}$$

Large-signal current gain α: This is the ratio of the negative of the collector-current increment from cutoff ($I_c = I_{co}$), to the emiiter-current change from cutoff ($I_E = 0$), e.g.,

$$\boxed{\alpha = \frac{-(I_c - I_{co})}{I_E - 0}}$$ The large-signal current gain of a C-B transistor.

23

α is not a constant, but vaires with emitter current I_E, V_{CB} and temperature.

A generalized transistor equation gives an expression for I_c in terms of any V_c and I_E:

$$I_c = -\alpha I_E + I_{co}(1 - e^{V_c/V_T})$$

A transistor as an amplifier, and parameter α':

A small voltage change ΔV_i between emitter and base causes a relatively large emitter-current change

$$\Delta I_E. \ \Delta I_c = \alpha' \cdot \Delta I_E, \quad \Delta V_L = -R_L \, \Delta I_c = -\alpha' \cdot R_L \cdot \Delta I_E.$$

If the dynamic resistance of the emitter junction is r_e, then

$$\Delta V_i = r_e \cdot \Delta I_E, \quad \text{and} \quad A = \alpha' \cdot R_L(\Delta I_E/r_e)\Delta I_E = \frac{-\alpha' \cdot R_L}{r_e}$$

The parameter α':

α' = The negative of the small-signal Short-circuit current transfer ratio $\equiv \dfrac{\Delta I_c}{\Delta I_E}\bigg|_{V_{CB}}$

$\alpha' = -\alpha$ (assuming α is independent of I_E)

3.1.4 TRANSISTOR CONFIGURATION

The CB configuration:

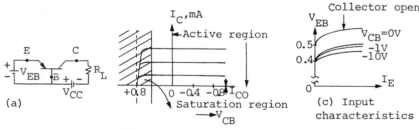

(a)

(b) Output characteristics

(c) Input characteristics

24

The common-emitter configuration:

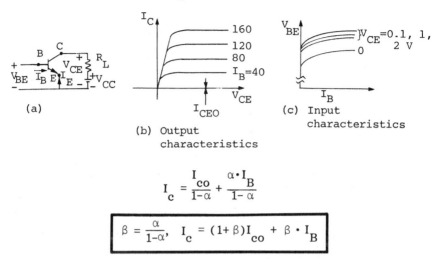

(a)

(b) Output
    characteristics

(c) Input
    characteristics

$$I_c = \frac{I_{co}}{1-\alpha} + \frac{\alpha \cdot I_B}{1-\alpha}$$

$$\beta = \frac{\alpha}{1-\alpha}, \quad I_c = (1+\beta)I_{co} + \beta \cdot I_B$$

CE cutoff currents – A transistor is in cutoff if $I_E = 0$ and $I_c = I_{co}$. It is not in cutoff if the base is open-circuited.

Common-emitter current gain:

Large-signal current-gain $\quad \beta = \dfrac{I_c - I_{CBo}}{I_B - (-I_{CBo})}$

DC current gain $h_{FE} = \beta_{dc} = \dfrac{I_C}{I_B}$

Small-signal current gain $\quad h_{fe} = \beta' = \left.\dfrac{\Delta I_c}{\Delta I_B}\right|_{V_{CE}}$

3.2 THE JUNCTION TRANSISTOR: SMALL-SIGNAL MODELS

3.2.1 THE HYBRID-Pi MODEL

It is useful to predict high-frequency performance.

Model:

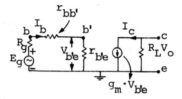

b,e,c: external terminals

$C_{b'c}$ is the depletion-region capacitance of the collector-base junction. C_{be} is an equivalent capacitance that accounts for a reduction in gain and in increase in phase shift at higher frequency.

Simplified circuit valid at low frequencies:

$$\frac{V_0}{V_{b'e}} = -g_m \cdot R_L \text{ (which is equivalent to } A_v \text{ if } r_{bb'} < r_{b'e})$$

Beta cutoff frequency f_β : At this frequency the magnitude of h_{fe} has decreased 3dB from its mid-frequency value.

$$h_{fe} = \frac{I_c}{I_b} = \frac{g_m \cdot r_{be}}{1 + j\omega/\omega\beta}$$

$$\omega_\beta = 1/r_{b'e}(c_{b'e} + c_{b'c})$$

$$f_\beta = \frac{1}{2\pi \cdot r_{b'e} \cdot c_{b'e}} \quad \text{(since } c_{b'e} \gg c_{b'c}\text{)}$$

$$f_T = h_{fe} \cdot f_\beta = \frac{g_m}{2\pi \cdot c_{b'e}}$$

3.2.2 h-PARAMETERS

h-Parameters for Each Transistor Configuration in Terms of the Other Two.

		Common base	Common collector
Common emitter	h_{ie}	$\dfrac{h_{ib}}{h_{fb} + 1}$	h_{ic}
	h_{fe}	$-\dfrac{h_{fb}}{h_{fb} + 1}$	$-(h_{fc} + 1)$
	h_{oe}	$\dfrac{h_{ob}}{h_{fb} + 1}$	h_{oc}
	h_{re}	$\dfrac{h_{ib}h_{ob}}{1 + h_{fb}} - h_{rb}$	$1 - h_{rc}$

		Common emitter	Common collector
Common base	h_{ib}	$\dfrac{h_{ie}}{h_{fe} + 1}$	$-\dfrac{h_{ic}}{h_{fc}}$
	h_{fb}	$\dfrac{-h_{fe}}{h_{fe} + 1}$	$-\dfrac{h_{fc} + 1}{h_{fc}}$
	h_{ob}	$\dfrac{h_{oe}}{h_{fe} + 1}$	$-\dfrac{h_{oc}}{h_{fc}}$
	h_{rb}	$\dfrac{h_{ie}h_{oe}}{h_{fe} + 1} - h_{re}$	$\dfrac{-h_{ic}h_{oc}}{h_{fc}} + h_{rc} - 1$

		Common emitter	Common base
	h_{ic}	h_{ie}	$\dfrac{h_{ib}}{1 + h_{fb}}$
Common collector	h_{fc}	$-(1+h_{fe})$	$\dfrac{-1}{1 + h_{fb}}$
	h_{oc}	h_{oe}	$\dfrac{h_{ob}}{1 + h_{fb}}$
	h_{rc}	$1-h_{re} \cong 1(\text{since } h_{re} <\!<1)$	$\dfrac{1-h_{ib}h_{ob}}{1 + h_{fb}}$

3.2.3 THE CONCEPT OF BIAS STABILITY

Q-point variation due to uncertainty in β :

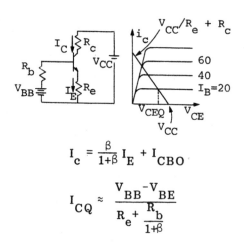

$$I_c = \frac{\beta}{1+\beta} I_E + I_{CBO}$$

$$I_{CQ} \approx \frac{V_{BB} - V_{BE}}{R_e + \dfrac{R_b}{1+\beta}}$$

$$I_{CQ} \approx \frac{V_{BB} - V_{BE}}{R_e} \approx \frac{V_{BB} - 0.7}{R_e}$$

iff $R_b <\!< \beta R_e$

28

The effect of temperature on the Q-point:

$$\Delta I_{CQ} = \frac{K \cdot \Delta T}{R_e} + \left(1 + \frac{R_b}{R_e}\right) I_{CBO1}(e^{K \cdot \Delta T} - 1)$$

Stability factor:

$$S_I = \frac{\Delta I_{CQ}}{\Delta I_{CBO}}, \quad S_V = \frac{\Delta I_{CQ}}{\Delta V_{BE}}, \quad S_\beta = \frac{\Delta I_{CQ}}{\Delta \beta}$$

for the common-emitter amplifier:

$$S_I \cong 1 + \left(\frac{R_b}{R_e}\right), \quad S_V \cong -1/R_e, \quad \text{and} \quad S_\beta = \frac{I_{CQ1}}{\beta_1}\left(\frac{R_b + R_e}{R_b + (1+\beta_2)Re}\right)$$

where $\beta_1, \beta_2, I_{CQ1}$ and I_{CQ2} are the lower and upper limits.

Temperature compensation using diode biasing:

Single diode compensations:

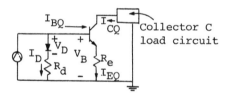

Collector C
load circuit

This compensation reduces the base-emitter voltage variation.

The diode is such that: $\dfrac{\Delta V_D}{\Delta T} = \dfrac{\Delta V_{BE}}{\Delta T}$.

$$I_{BB} = I_D + \frac{I_{EQ}}{1+\beta} = \text{constant}$$

$$V_B = V_D + I_D \cdot R_d = V_{BEQ} + I_{EQ} \cdot R_e$$

$$I_{EQ} = \frac{V_D - V_{BEQ} + I_D R_d}{R_e + R_d / (1 + \beta)} \quad \text{and} \quad \frac{\Delta I_{EQ}}{\Delta T} = 0$$

Two-diode compensation:

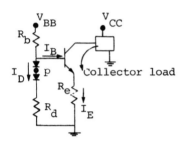

The quiescent emitter current is independent of variations of temperature if $R_b = R_d$.

$$I_{EQ} = \frac{(V_{BB} \cdot R_d + 2V_D \cdot R_b) / (R_b + R_d) - V_{BEQ}}{R_e}$$

CHAPTER 4

POWER SUPPLIES

4.1 DIODE RECTIFIERS

4.1.1 HALF-WAVE RECTIFIER

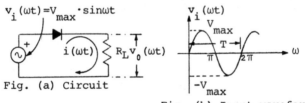

$v_i(\omega t) = V_{max} \cdot \sin\omega t$

i(ωt)

$R_L v_0(\omega t)$

Fig. (a) Circuit

$v_i(\omega t)$

V_{max}

T

π 2π ω

$-V_{max}$

Fig. (b) Input waveform

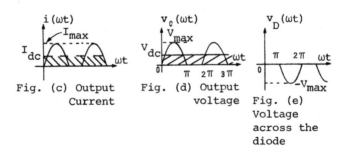

i(ωt)

I_{max}

I_{dc} ωt

Fig. (c) Output
Current

$v_c(\omega t)$

V_{max}

V_{dc} ωt

π 2π 3π

Fig. (d) Output
voltage

$v_D(\omega t)$

π 2π ωt

$-V_{max}$

Fig. (e)
Voltage
across the
diode

$$I_{dc} = \frac{I_{max}}{\pi} = 0.318\, I_{max}$$

$$v_{dc} = 0.318\, I_{max} \cdot R_L = 0.318\, V_{max}$$

4.1.2 FULL-WAVE RECTIFIER

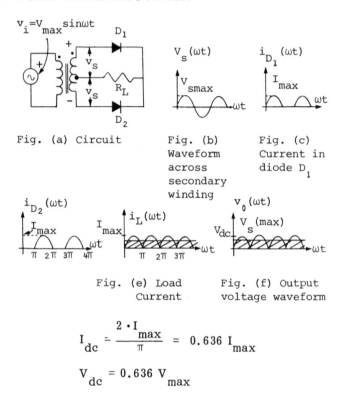

Fig. (a) Circuit

Fig. (b)
Waveform
across
secondary
winding

Fig. (c)
Current in
diode D_1

Fig. (e) Load
Current

Fig. (f) Output
voltage waveform

$$I_{dc} = \frac{2 \cdot I_{max}}{\pi} = 0.636\, I_{max}$$

$$V_{dc} = 0.636\, V_{max}$$

4.1.3 PIV RATING

The PIV rating indicates the voltage which a rectifier diode can withstand in a reverse bias condition.

$$PIV = 2 \cdot V_{max} \quad \text{- for conventional full-wave rectifier}$$

$$= V_{max} \quad \text{- for the bridge rectifier}$$

4.1.4 RIPPLE FACTOR

Ripple factor (RF) = $\dfrac{\text{rms value of the ac component}}{\text{dc value of the waveform}}$

$i_{ac} = i_L - i_{dc}$ (i_L = the rectified load current)

$I_{ac(rms)} = [I_{L(rms)}^2 - I_{dc}^2]^{\frac{1}{2}}$

$RF = \left[\left(\dfrac{V_{L\,rms}}{V_{dc}} \right)^2 - 1 \right]^{\frac{1}{2}}$

$RF_{Halfwave} = 1.21$ $\left(V_{Lrms} = \dfrac{V_{max}}{2}$ and $V_{dc} = \dfrac{V_{max}}{\pi} \right)$

$RF_{Fullwave} = 0.482$ $\left[V_{L(max)} = \dfrac{V_{max}}{\sqrt{2}} = 0.707\, V_{max} \right]$

$RF_{Bridge-rectifier} = 0.482$ $\left[I_{L(max)} = \dfrac{I_{max}}{\sqrt{2}} = 0.707\, I_{max} \right]$

4.1.5 RECTIFIER EFFICIENCY

$\eta_R = \dfrac{P_{L(dc)}}{P_{i(dc)}}$

For the half-wave rectifier:

$P_{L(dc)} = I_{dc} \cdot V_{dc} = \left(\dfrac{I_{max}}{\pi} \right)^2 \cdot R_L$

$P_{i(ac)} = I_{rms}^2 \cdot R_L = \left(\dfrac{I_{max}}{2} \right)^2 \cdot R_L$

$\eta_R = 0.406$

For conventional and bridge rectifier:

$\eta_R = 0.812$

4.1.6 COMPARISON OF RECTIFIERS

Rectifier	V_{dc}	RF	f_{output}	η_R	PIV.
Half-wave	$0.318V_{s(max)}$	1.21	f_{input}	40.6%	$V_{s(max)}$
Full-wave	$0.636V_{s(max)}$	0.482	$2f_{input}$	81.2%	$2V_{s(max)}$
Bridge	$0.636V_{s(max)}$	0.482	$2f_{input}$	81.2%	$V_{s(max)}$

% load regulation:

$$\%LR = \frac{V_{NL} - V_{FL}}{V_{FL}} \times 100\%$$

V_{NL} = No-load dc voltage

V_{FL} = Full-load dc voltage

A high % LR is poor regulation.

4.2 FILTERS

4.2.1 SHUNT-CAPACITANCE FILTER

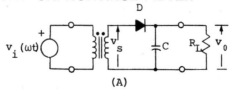

(A)

34

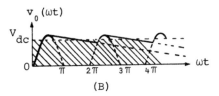

(B)

Half-wave rectifier with shunt-capacitance
filter. (A) Circuit. (B) Output waveform.

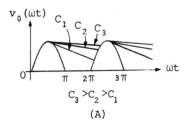

$C_3 > C_2 > C_1$

(A)

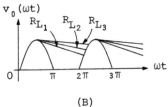

(B)

Output of Half-wave rectifier with
(A) C and (B) R_L vary.

Half-Wave Rectifier

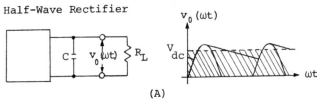

(A)

Full-Wave Rectifier

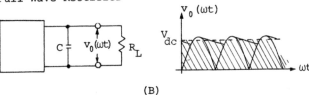

(B)

The same value of capacitance used with a full-
wave rectifier in (B) produces a higher dc output than when used with a half-wave rectifier
in (A).

35

The dc load voltage of a full-wave rectifier with a shunt-capacitance filter is given as:

$$V_{dc} = \frac{V_{s(max)}}{1 + \dfrac{1}{4f \ R_L \ C}}$$

$$RF = \frac{1}{4\sqrt{3} \cdot f \cdot C \cdot R_L}$$

4.2.2 Pi-FILTER

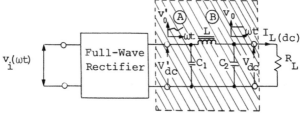

Full-wave rectifier and pi filter.

$$V_{dc} = V'_{dc} - I_{dc} \cdot r_L = \frac{V'_{dc} \cdot R_L}{R_L + r_L}$$

$$RF = \sqrt{2} / \ \omega^3 \cdot R_L \cdot C_1 \cdot C_2 \cdot L$$

4.2.3 RC FILTER

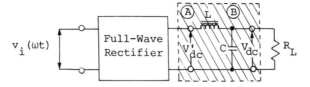

Full-wave rectifier and L-section (choke-input) filter.

$$V_{dc} = \frac{V'_{dc} \cdot R_L}{R_L + R}$$

4.2.4 L-SECTION FILTER

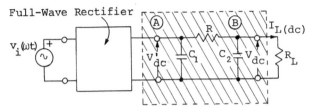

Full-wave rectifier and RC filter.

This filter is used with high-load current circuits.

4.2.5 VOLTAGE MULTIPLIER

The voltage multiplier is used when a high dc voltage with extremely light loading is required.

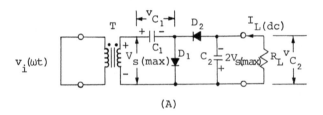

(A)

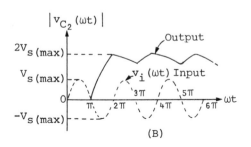

(B)

$$V_{c2} = 2V_{s(max)}$$

4.2.6 COMPARISON OF FILTERS

Filter	V_{dc}	RF
Shunt-Capacitance Rectifier FW 	$\dfrac{V_{s(max)}}{1 + 1/(4fR_LC)}$ For 60 Hz: $\dfrac{V_{s(max)}}{0.00417/R_LC}$	$\dfrac{1}{4\sqrt{3}fR_LC}$ For 60 Hz: $\dfrac{2.41 \times 10^{-3}}{R_LC}$
Pi Rectifier FW 	$\dfrac{V_{s(max)}R_L/(R_L+r_L)}{1 + 1/(4fR_LC_1)}$ For 60 Hz: $\dfrac{V_{s(max)}R_L/(R_L+r_L)}{1 + 0.00417/R_LC}$	$\dfrac{\sqrt{2}}{\omega^3 C_1C_2LR_L}$ For 60 Hz: $\dfrac{0.026}{C_1C_2R_LL} \times 10^{-6}$
RC Rectifier FW 	$\dfrac{V_{s(max)}R_L/(R_L+R)}{1 + 1/(4fR_LC_1)}$ For 60 Hz: $\dfrac{V_{s(max)}R_L/(R_L+R)}{1 + 0.00417/R_LC}$	$\dfrac{\sqrt{2}}{\omega^2 C_1C_2R_LR}$ For 60 Hz: $\dfrac{9.95}{C_1C_2R_LR} \times 10^{-6}$
L-Section Rectifier FW 	$0.636V_{s(max)}$ For 60 Hz: $0.636V_{s(max)}$	$\dfrac{0.118}{(\omega^2 LC)}$ For 60 Hz: $\dfrac{0.83}{LC} \times 10^{-6}$

CHAPTER 5

MULTITRANSISTOR CIRCUITS

5.1 THE DIFFERENCE AMPLIFIER

5.1.1 BASIC DIFFERENCE AMPLIFIER

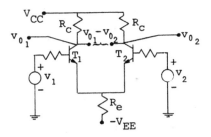

The Differential-mode or difference-mode input voltage
$v_d = v_2 - v_1$.

The common-mode input voltage $V_a = \frac{V_2 + V_1}{2}$.

$$v_2 = v_a + \frac{v_d}{2}$$

$$v_1 = v_a - \frac{v_d}{2}$$

V_d is the desired signal and is amplified while V_a is rejected.

5.1.2 Q-POINT ANALYSIS

The differential-mode input is assumed to be zero. $(V_1 = V_2)$

The equivalent circuit for T_1 or T_2 when $v_1 = v_2 = v_a$:

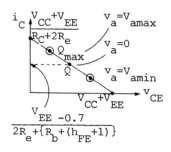

$$v_{E_1} = v_{E_2} = (i_{E_1} + i_{E_2})R_e - V_{EE}$$

$$= i_E(2R_e) - V_{EE} \quad (\text{when } i_{E_1} = i_{E_2} = i_E)$$

The load-line equation for $v_a = v_1 = v_2$ is

$$v_{CE} = V_{CC} - i_C R_C - i_E(2R_e) + V_{EE}$$

$$\approx V_{CC} + V_{EE} - i_C(R_C + 2R_e)$$

$$i_C \approx \frac{v_a + V_{EE} - 0.7}{2R_e + [R_b \div (h_{FE}+1)]}, \quad (V_{BE} = 0.7))$$

5.1.3 COMMON-MODE LOAD LINE

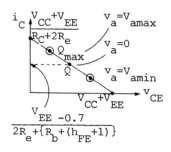

The common-mode input $v_a = 0$, for Q.

$$Q = Q_{max} \quad \text{when} \quad v_a = v_{amax};$$

$$Q = Q_{min} \quad \text{when} \quad v_a = v_{amin}.$$

In both cases, $v_d = 0$.

The individual collector voltages v_{01} and v_{02} will vary with variations in v_a.

Difference-mode load-line equations

These equations determine the effect of a non-zero difference-mode input. ($v_2 = -v_1 = \frac{v_d}{2}$, $v_a = 0$ and Q is as shown in the previous figure.)

$$\Delta v_{CE_1} = -R_c \cdot \Delta i_{c_1}, \text{ using small-signal notation}$$

$$\boxed{v_{ce_1} = -R_c \cdot i_{c_1}} \quad , \quad \Delta v_{CE_2} = -R_c \cdot \Delta i_{c_2},$$

$$\text{or} \quad \boxed{v_{ce_2} = -R_c \cdot i_{c_2}}$$

These are "Difference-mode load line" equations, such that $v_{E_1} = v_{E_2}$.

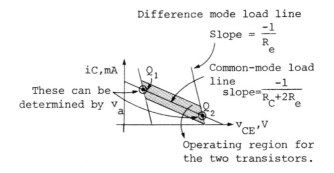

Difference mode load line
Slope $= \dfrac{-1}{R_e}$

iC, mA

These can be determined by v_a

Common-mode load line slope $= \dfrac{-1}{R_C + 2R_e}$

v_{CE}, V

Operating region for the two transistors.

41

Small-signal analysis:

Equivalent circuit with all components reflected into emmiter.

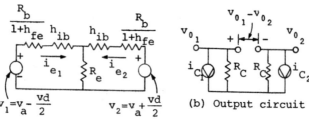

(a) Input circuit

(b) Output circuit

Circuit used to calculate i_a and i_d:

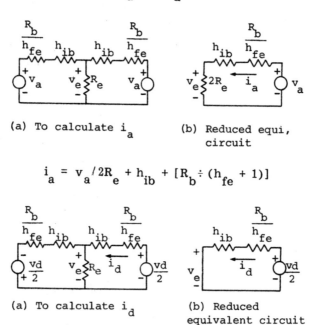

(a) To calculate i_a

(b) Reduced equi, circuit

$$i_a = v_a / 2R_e + h_{ib} + [R_b \div (h_{fe} + 1)]$$

(a) To calculate i_d

(b) Reduced equivalent circuit

$$i_d = \frac{v_d / 2}{h_{ib} + [R_b \div (1 + h_{fe})]}$$

Assuming $i_c \approx i_e$, and $v_{01} - v_{02} = \dfrac{R_c}{h_{ib} + [R_b \div (1 + h_{fe})]} \cdot v_d$

42

Common-mode Rejection Ratio:

A_d = The difference-mode gain = $\dfrac{R_c/2}{h_{ib}+[R_b \div (1+h_{fe})]}$

A_a = The common-mode gain = $\dfrac{R_c}{2R_e+h_{ib}+[R_b \div (1+h_{fe})]}$

$v_{01} = A_d \cdot V_d - A_a \cdot V_a$ and $v_{02} = -A_d \cdot v_d - A_a \cdot v_a$.

Common-mode rejection ratio:

$$\text{"CMRR"} = \frac{A_d}{A_a} \cong \frac{R_e}{h_{ib}+[R_b/h_{fe})}$$

If "CMRR" $>> \dfrac{V_a}{V_d}$, then the output voltage is proportional to V_d.

5.1.4 DIFFERENCE AMPLIFIER WITH CONSTANT CURRENT SOURCE

$\text{CMRR} = \dfrac{R_e}{(V_T/I_{EQ})+(R_b/h_{fe})}$, if $\dfrac{R_b}{h_{fe}}$ is small,

$$\text{CMRR} < \frac{R_e \cdot I_{EQ}}{V_T}$$

CMRR can be increased only by increasing R_e.

In the previous circuit, R_e is replaced by another transistor which is a constant-current source.

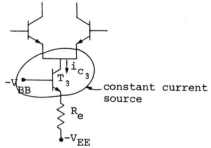

constant current source

Quiescent operation:

$$I_{CQ_3} \approx \frac{V_{EE} - V_{BB} - 0.7}{R_e}$$

(I_{CQ_3} is constant as long as T_3 does not saturate.)

The condition for keeping T_3 in the linear region is:

$$v_{CE} > V_T \left[2.2 + \ln \left(\frac{h_{fe}}{h_{fe}} \right) \right]$$

Load-lines that describe circuit operation:

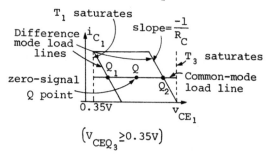

$$\left(V_{CEQ_3} \geq 0.35V \right)$$

Small-signal operation:

Equivalent circuit looking into the bases

The input impedance R_i between the bases of T_1 and T_2 is:

$$R_i = 2 \cdot h_{ie}.$$

44

5.1.5 DIFFERENCE AMPLIFIER WITH EMITTER RESISTORS FOR BALANCE

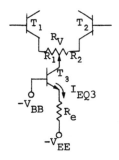

Difference amplifier with
balance control R_U.

To compensate for different h_{fe1} and h_{fe2} , R_v is used when T_1 and T_2 have different characteristics.

The condition which ensures that the emitter currents of T_1 and T_2 are the same is:

$$R_2 - R_1 = R_b \left(\frac{1}{h_{fe1}} - \frac{1}{h_{fe2}} \right) .$$

R_v results in symmetrical operation but it causes a loss in the current gain.

$$A_d = \frac{R_c}{R_b \left(\frac{1}{h_{fe1}} + \frac{1}{h_{fe2}} \right) + 2h_{ib} + Rv} ,$$

if

$$(R_e)_{eff} \approx \frac{1}{h_{ob3}} \gg h_{ib1} + \frac{R_b}{h_{fe1}} + R_1$$

5.2 THE DARLINGTON AMPLIFIER

The Darlington amplifier is used to provide increased input impedance and a very high current gain $(h_{fe_1} \cdot h_{fe_2})$.

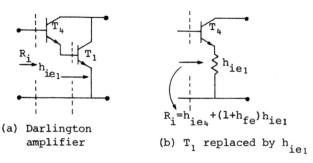

Basic Darlington amplifier

Current gain $A_i = \alpha(1+h_{fe})^2 + \alpha(1+h_{fe}) = \alpha(1+h_{fe})(h_{fe}+2)$

$$\approx h_{fe}^2 \quad \text{(assuming identical transistors)}$$

$$R_i = h_{ie_4} + (1+h_{fe})h_{ie_1}$$

$$= \frac{(1+h_{fe})V_T}{I_{EQ_4}} + \frac{(1+h_{fe})^2 \cdot V_T}{I_{EQ_1}}$$

$$= 2(1+h_{fe})h_{ie_1} = 2 \cdot h_{ie_4}$$

Input impedance:

(a) Darlington amplifier

(b) T_1 replaced by h_{ie_1}

$$R_i = h_{ie_4} + (1+h_{fe})h_{ie_1}$$

In this case, emitter current in T_4 can be adjusted by setting R_b,

$$I_{EQ_4} = I_{BQ_1} + \frac{0.7}{R_b}$$

46

$$R_i = h_{ie_4} + (1+h_{fe})(R_b \| h_{ie_1})$$

$$i_0 = h_{fe} \cdot i_{b_4} + h_{fe}(1 + h_{fe})i_{b_4} \cdot \frac{R_b}{R_b + h_{ie_1}}$$

$$A_i = h_{fe}^2 \frac{R_b}{R_b + h_{ie}} + h_{fe} \cdot \left(1 + \frac{R_b}{R_b + h_{ie_1}}\right)$$

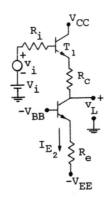

Darlington amplifier with bias resistor.

5.3 THE CASCADE AMPLIFIER

The Amplifier:

The cascade amplifier is used as a dc level shifter when

the voltage of interest consists of a small-signal ac component v_i and a fixed dc level V_i. v_i from the level shifter should have a dc level different from V_i. Typically this final output level is to be OV.

DC analysis:

T₁ is the emitter follower and T₂ acts as a constant current source. DC component of the output voltage, V_L:

$$V_i - \frac{R_i\, I_{E2}}{1+h_{fe}} - 0.7 - R_c \cdot I_{E2}$$

Small-signal analysis:

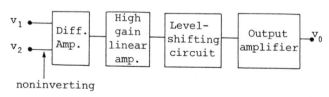

v_L (small-signal component of the output voltage) $\cong$ $\dfrac{v_i}{1 + \dfrac{(R_i \div h_{fe})+h_{ib}+R_c}{1\ /\ h_{ob2}}}$

Since $\dfrac{1}{h_{ob2}}$ is much larger than $\dfrac{R_i}{h_{fe}} + h_{ib} + R_c$, the load voltage $v_L \approx v_i$, while a negligible attenuation of the signal has resulted from the shift in dc level.

5.4 THE OP AMPLIFIER

Typical configuration:

CHAPTER 6

SMALL-SIGNAL, LOW-FREQUENCY ANALYSIS AND DESIGN

6.1 HYBRID PARAMETERS

6.1.1 GENERAL TWO-PORT NETWORK

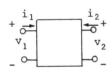

(a) Two-part network

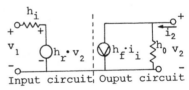

Input circuit Ouput circuit

(b) Equivalent circuit of two-part network.

6.1.2 HYBRID EQUATIONS

$$v_1 = h_i \cdot i_1 + h_r \cdot v_2$$

$$i_2 = h_f \cdot i_1 + h_0 \cdot v_2$$

$$v_1 = h_{11} \cdot i_1 + h_{12} \cdot v_2$$

$$i_2 = h_{21} \cdot i_1 + h_{22} \cdot v_2$$

49

6.1.3 TERMINAL DEFINITIONS FOR H-PARAMETERS

$h_i = \dfrac{v_1}{i_1}\bigg|_{v_2 = 0}$ = short-circuit input impedance

$h_r = \dfrac{v_1}{v_2}\bigg|_{i_1 = 0}$ = open-circuit reverse voltage gain

$h_f = \dfrac{i_2}{i_1}\bigg|_{v_2 = 0}$ = short-circuit forward current gain

$h_0 = \dfrac{i_2}{v_2}\bigg|_{i_1 = 0}$ = open-circuit output admittance

6.2 THE C-E CONFIGURATION

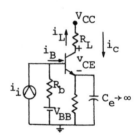

(a) Complete circuit

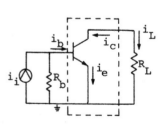

(b) Small-signal circuit

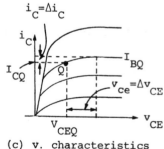

(c) v_i characteristics

$$h_{oe} = \left.\frac{i_c}{v_{ce}}\right|_{ib=0} = \left.\frac{\Delta i_c}{\Delta v_{ce}}\right|_{Q \text{ point}}$$

where i_c and v_{ce} are small variations about nominal operating point.

$$h_{oe} = \frac{h_{FE}}{h_{FC}} \cdot \frac{I_{CQ}}{V_T} \left[\frac{e^{V_{CE}/V_T}}{[e^{V_{CE}/V_T} + h_{FE}/h_{FC}]^2} \right]$$

$$\simeq \frac{h_{FE}}{h_{FC}} \frac{I_{CQ}}{V_T} e^{-V_{CE}/V_T}$$

$$h_{fe} = \left.\frac{i_c}{i_b}\right|_{Q \text{ point}} = \left.\frac{\Delta i_C}{\Delta i_B}\right|_{Q \text{ point}}$$

$$h_{ie} = \left.\frac{v_{be}}{i_b}\right|_{V_{CE}=0} = \left.\frac{v_{be}}{i_b}\right|_{Q \text{ point}}$$

$$= \frac{V_T}{I_{BQ}} \simeq h_{fe} \cdot \frac{V_T}{I_{CQ}} \simeq h_{fe} \cdot \frac{V_T}{I_{EQ}}$$

Equivalent circuit:

This diagram can be simplified by ignoring h_{oe} and h_{re}.

C-E amplifier equivalent circuit:

$$\frac{i_b}{i_i} = \frac{R_b}{R_b + h_{ie}}$$

$$A_i = \frac{i_L}{i_i} = \frac{-h_{fe}}{1 + h_{fe}[(25 \times 10^{-3})/I_{EQ}R_b]}$$

Requirement of high gain and stability:

$$h_{ie} = h_{fe} \cdot \frac{V_T}{I_{EQ}} \ll R_b \ll h_{fe} \cdot R_e$$

$$Z_i = \frac{R_b h_{ie}}{R_b + h_{ie}} \approx h_{ie} \quad (\text{if } R_b \gg h_{ie})$$

$$Z_0 = \frac{v_{ce}}{i_c}\Bigg|_{i_i = 0} = 1/h_{oe}$$

6.3 THE C-B CONFIGURATION

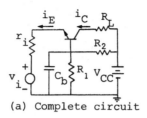

(a) Complete circuit

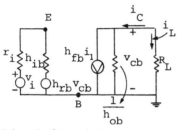

(b) Hybrid model

(c) Simplified circuit

For the hybrid model:

$$V_{eb} = h_{ib}i_1 + h_{rb} \cdot v_{cb} = h_{ib}(-i_e) + h_{rb} \cdot v_{cb}$$

$$i_c = h_{fb} \cdot i_1 + h_{ob} \cdot v_{cb} = h_{fb}(-i_e) + h_{ob} \cdot v_{cb}$$

$$h_{ib} = \left.\frac{v_{eb}}{-i_e}\right|_{v_{cb}=0} = \frac{V_T}{I_{EQ}} \approx \frac{h_{ie}}{1+h_{fe}}$$

$$h_{fb} = \left.\frac{i_c}{i_1}\right|_{v_{cb}=0} = \left.\frac{i_c}{-i_e}\right|_{v_{cb}=0} = -\alpha$$

$$h_{ob} = \left.\frac{i_c}{v_{cb}}\right|_{i_e=i_1=0}$$

53

Simplified equivalent circuit:

$$i_e = (1 + h_{fe})i_b = (i + h_{fe})\left(\frac{-v_{eb}}{h_{ie}}\right)$$

$$h_{ib} = \left.\frac{-v_{eb}}{i_e}\right|_{v_{cb}=0} = \frac{h_{ie}}{1+h_{fe}} \approx \frac{h_{ie}}{h_{fe}}$$

$$h_{fb} \text{ (short-circuit current gain)} = \left.\frac{i_c}{-i_e}\right|_{v_{cb}=0} = \frac{-h_{fe}}{1+h_{fe}} \doteq -1$$

$$h_{ob} \approx \frac{h_{oe}}{h_{fe}}$$

Modified circuits:

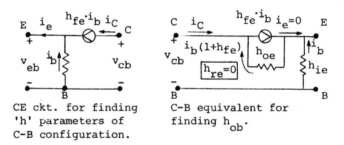

CE ckt. for finding 'h' parameters of C-B configuration.

C-B equivalent for finding h_{ob}.

To find the CB parameters h_{ob}, h_{fb}, and h_{ib}, divide the corresponding CE parameters by $1 + h_{fe}$.

6.4 THE C-C (EMITTER-FOLLOWER) CONFIGURATION

Characteristics:

A) A voltage gain slightly less than unity;

B) A high input impedance, and

C) A low output impedance.

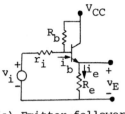

(a) Emitter-follower.

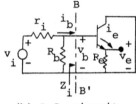

(b) A.C. circuit

(c) Equivalent circuit

$$A_v = \frac{v_e}{v_i} = \frac{R_b}{r_i + R_b}\left[\frac{1}{1+[h_{ie}+(r_i\|R_b)]/[(1+h_{fe})R_e]}\right]$$

$$Z_i = h_{ie} + (1+h_{fe})R_e$$

$$Z_0 = h_{ib} + \frac{r_i}{1+h_{fe}}$$

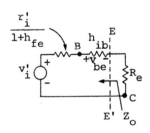

6.5 SIGNIFICANT PARAMETERS

	Configuration		
	CE	EF (CC)	CB
Gain	$A_i \approx -h_{fe}$	$A_v \approx 1$	$A_i \approx -h_{fb} = \dfrac{-h_{fe}}{1+h_{fe}}$
Input impedance	$h_{ie} = \dfrac{(25\times10^{-3})h_{fe}}{I_{EQ}}$	$z_i = h_{ie} + (h_{fe}+1)R_e$	$h_{ib} = \dfrac{h_{ie}}{1+h_{fe}}$
Output impedance	$\dfrac{1}{h_{oe}} > 10^4\,\Omega$	$z_o \approx h_{ib} + \dfrac{r_i'}{h_{fe}+1}$	$\dfrac{1}{h_{ob}} = \dfrac{1+h_{fe}}{h_{oe}}$
Simplest equivalent circuit		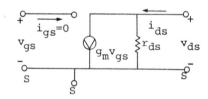	

6.6 SMALL-SIGNAL EQUIVALENT CIRCUIT OF THE FET

6.6.1 EQUIVALENT CIRCUIT

6.6.2 TRANSCONDUCTANCE

$$g_m = \left.\frac{\partial i_{DS}}{\partial v_{GS}}\right|_{Q\ point}$$

For the MOSFET:

$$i_{DS} = k_n(V_{GS}-V_T)^2, \text{ and } g_m = 2k_n(V_{GS}-V_T)\Big|_{V_{GSQ}}$$

$$= 2\sqrt{k_n \cdot I_{DSQ}}$$

g_m of a FET is analogous to $1/h_{ib}$ in BJT:

$$\left[\frac{1}{h_{ib}}\right]_{BJT} \gg (g_m)_{FET}$$

6.6.3 DRAIN SOURCE RESISTANCE

$$r_{ds} = \frac{\partial V_{DS}}{\partial i_{DS}}\Big|_{Q \text{ point}} \quad \alpha \ \frac{1}{I_{DQ}}$$

The drain source resistance is analogous to h_{oe} of the transistor.

6.6.4 AMPLIFICATION FACTOR

$$\mu = \frac{-\partial v_{DS}}{\partial v_{GS}}\Big|_{Q \text{ point}} = g_m \cdot r_{ds}$$

Equivalent Model:

6.7 THE COMMON-SOURCE VOLTAGE AMPLIFIER

$$A_v = \frac{V_L}{V_i} = -g_m(R_L \| Z_0)\left[\frac{1}{1+[r_i \div (R_3+(R_1\|R_2)]}\right]$$

with $r_i \ll R_3 + (R_1 \| R_2)$ and $R_L \ll Z_0$, $A_V \doteq -g_m R_L$

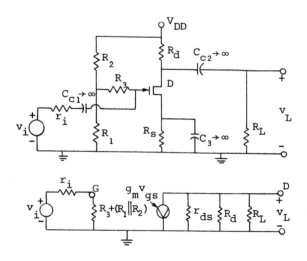

6.8 THE COMMON-DRAIN VOLTAGE AMPLIFIER (THE SOURCE FOLLOWER)

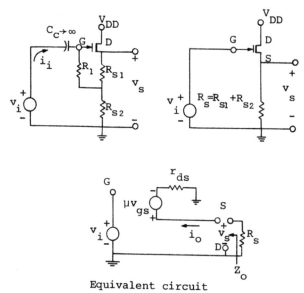

Equivalent circuit

$$Z_0 = \left.\frac{v_s}{i}\right|_{v_i=0} \text{ (as seen by } R_s) = \frac{r_{ds}}{1+\mu} \approx 1/g_m$$

$$\text{(when } \mu = g_m \cdot r_{ds} \gg 1)$$

A'_v (open-circuit voltage gain) $= \frac{\mu}{1+\mu} \cong 1$, when $\mu \gg 1$

$$A_v = \frac{v_s}{v_i} = \frac{\mu}{1+\mu}\left[\frac{g_m \cdot R_s}{1+g_m \cdot R_s}\right]$$

$$Z_i = \frac{v_g}{i_i} \cong \frac{R_1}{1 - \left(\dfrac{v_s}{v_g}\right)\left[\dfrac{R_{s2}}{R_{s1}+R_{s2}}\right]}$$

Z_{02} (looking into the source) $\cong \dfrac{R_s(r_{ds}+R_s)}{r_{ds}+(2+\mu)R_s}$

$$\cong \frac{r_{ds}+R_s}{\mu}$$

Equivalent circuit for Z_i:

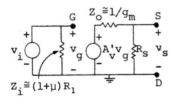

Assuming $R_{s2} \gg R_{s1}$,

$$Z_i \cong (1+\mu)R_1$$

59

6.9 THE PHASE-SPLITTING CIRCUIT

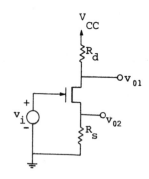

$$v_{01} = \frac{-\mu \cdot R_d}{(1+\mu)R_s + r_{ds} + R_d} \, v_i$$

$$v_{02} = \frac{(\mu/1+\mu)R_S}{R_s + \left(\dfrac{r_{ds}}{1+\mu}\right) + \dfrac{R_d}{1+\mu}} \, v_i$$

Z_{01} (looking into the source) $\cong \dfrac{R_s[r_{ds} + (1+\mu)R_s]}{r_{ds} + (2+\mu)R_s}$

$$\cong R_s$$

CHAPTER 7

AUDIO-FREQUENCY LINEAR POWER AMPLIFIERS

7.1 THE CLASS A COMMON-EMITTER POWER AMPLIFIER

The power amplifiers are classified according to the portion of the input sine wave cycle during which the load current flows.

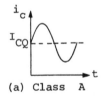

(a) Class A

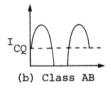

(b) Class AB

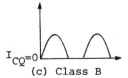

(c) Class B

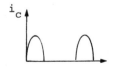

(d) Class C current flows for less than one-half cycle

Q-point placement:

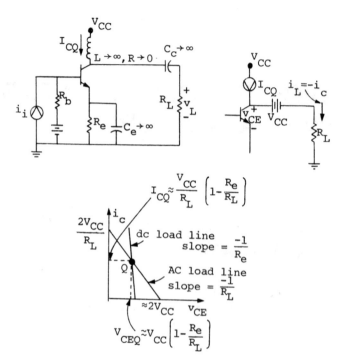

DC-load-line equation:

$$V_{CC} = v_{CE} + i_C \cdot R_e$$

AC-load-line equation:

$$v_{ce} = -i_c R_L = i_L \cdot R_L$$

or

$$i_C - I_{CQ} = \frac{-1}{R_L} (v_{CE} - V_{CEQ})$$

To place "Q" for the maximum symmetrical swing

$$I_{CQ} = \frac{V_{CC}}{R_{ac} + R_{dc}}$$

$$\approx V_{CC}/R_L$$

$$V_{CEQ} = V_{CC} \frac{R_L}{R_L + R_e} = \frac{V_{CC}}{1 + (R_e / R_L)}$$

$$\approx V_{CC} \text{ (because } R_L \gg R_e)$$

Power calculations:

$$i_C = I_{CQ} + i_c = \frac{V_{CC}}{R_L} + i_c$$

$$i_L = -i_c$$

$$i_{supply} = i_L + i_C = I_{CQ} = \frac{V_{CC}}{R_L}$$

$$v_{CE} = V_{CC} - i_c R_L, \quad v_L = i_L R_L = -i_c R_L$$

For a sinusoidal signal current:

$$i_i = I_{im} \cdot \sin \omega t, \quad i_c = I_{cm} \cdot \sin \omega t$$

Supplied power $P_{CC} = V_{CC} \cdot I_{CQ} \approx \dfrac{V_{CC}^2}{R_L}$

Power transferred to load:

$$P_L = \frac{I_{LM}^2 \cdot R_L}{2} = \frac{I_{cm}^2 \cdot R_L}{2}$$

(since $i_L = -i_c$, $I_{LM} = -I_{cm}$).

$$P_{Lmax} = \frac{I_{CQ}^2 \cdot R_L}{2} = \frac{V_{CC}^2}{2R_L} \quad \text{(when } I_{cm} = I_{CQ}).$$

Collector dissipation:

$$P_C = P_{CC} - P_L = \frac{V_{CC}^2}{R_L} - \frac{I_{cm}^2 \cdot R_L}{2}$$

$$P_{Cmin} = V_{CC}^2 / 2R_L \quad \text{and} \quad P_{Cmax} = \frac{V_{CC}^2}{R_L} = V_{CEQ} I_{CQ}$$

Efficiency:

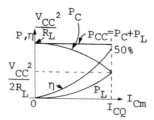

Variation of power and efficiency
with collector current

$$\eta = \frac{P_L}{P_{CC}} = \frac{I_{cm}^2 (R_L/2)}{V_{CC} \cdot I_{CQ}} = \frac{1}{2}\left(\frac{I_{cm}}{I_{CQ}}\right)^2$$

$\eta_{max} = 50\%$

Figure of merit $= \dfrac{P_{Cmax}}{P_{L\,max}} = 2.$

The maximum-dissipation hyperbola:

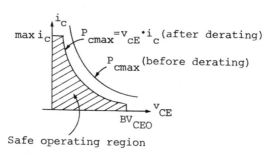

Safe operating region

For the safe operation, the Q-point must lie on or below the hyperbola $v_{CE} \cdot i_c = P_{c_1 max}$.

This hyperbola represents the locus of all operating points at which the collector dissipation is exactly P_{cmax}.

The ac load line, with slope $\dfrac{-1}{R_L}$, must pass through the Q-point, intersect the v_{CE} at a voltage less than BV_{CEO} and intersect the i_c axis at a current less than max i_c, i.e.,

64

$$2\,V_{CC} \leq BV_{CEO},$$

$$2\,I_{CQ} \leq i_{cmax}.$$

$$\text{Slope of hyperbola}\,\bigg|_{Q\text{ point}} = \frac{\partial i_c}{\partial v_{CE}} = \frac{-I_{CQ}}{V_{CEQ}} = \frac{-1}{R_L}$$

7.2 THE TRANSFORMER-COUPLED AMPLIFIER

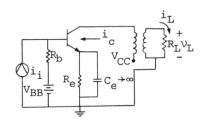

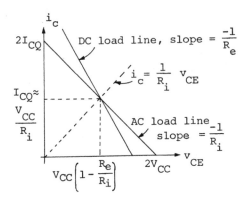

Load lines:

DC Load line: $\quad V_{CC} = v_{CE} + i_E \cdot R_e \approx v_{CE} + i_c R_e$

AC load line:
$$\text{Slope} = \frac{i_c}{v_{ce}} = \frac{-1}{R_L'} \quad (\text{where } R_L' = N^2 R_L).$$

$$I_{CQ} \approx \frac{V_{CC}}{R_L'} \left(1 - \frac{R_e}{R_L'} \right), \quad V_{CEQ} \approx V_{CC} \left(1 - \frac{R_e}{R_L'} \right)$$

Power calculations:

The signal i_L is sinusoidal; thus,

$$i_c = I_{cm} \cdot \sin \omega t.$$

Supplied power: $P_{CC} = V_{CC} \cdot I_{CQ} = \dfrac{V_{CC}^2}{R_L'}$

Load power:

$$P_L = \frac{I_{LM}^2}{2} \quad R_L = \frac{I_{cm}^2}{2} R_L', \quad (I_{LM} = N \cdot I_{CM})$$

$$P_{Lmax} = \frac{V_{CC}^2}{2R_L'}$$

Collector dissipation:

$$P_C = \frac{V_{CC}^2}{R_L'} - \frac{I_{cm}^2}{2} R_L'$$

$$P_{cmax} = \frac{V_{CC}^2}{R_L'} = V_{CEQ} \cdot I_{CQ}$$

Efficiency:

$$\eta = \frac{1}{2} \left(\frac{I_{cm}}{I_{CQ}} \right)^2$$

$$\eta_{max} = 50\%$$

Fig. of merit $= \dfrac{P_{cmax}}{P_{Lmax}} = 2.$

7.3 CLASS B PUSH-PULL POWER AMPLIFIERS

Circuit:

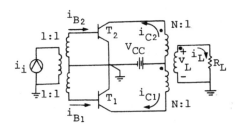

Waveforms:

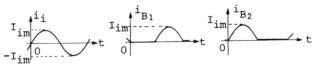

(a) Input current (b) Base current in T_1 (c) in T_2

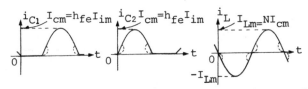

(d) Collector current in T_1 (e) Collector current in T_2 (f) Load current

Load-line determination:

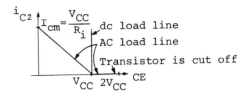

DC load line: $v_{CE} = V_{CC}$

AC load line:
$$\text{Slope} = \frac{i_c}{CE} = \frac{-1}{R'_L}$$

The maximum value of both i_{c_1} and i_{c_2} is $I_{cm} = \dfrac{V_{CC}}{R'_L}$.

Power calculations:

$$i_i = I_{im} \cdot \sin \omega t$$

Supplied power $= P_{CC} = \dfrac{2}{\pi} \cdot V_{CC} \cdot I_{cm}$

$$P_{CC(max)} = \frac{2}{\pi} V_{CC} \cdot \frac{V_{CC}}{R'_L} = \frac{2 \, V_{CC}^2}{\pi \cdot R'_L}$$

$$P_L = \frac{1}{2} I_{cm}^2 \cdot R'_L, \qquad P_{L,\,max} = \frac{V_{CC}^2}{2 R'_L}$$

Power dissipated in the collector:

$$2 P_C = \frac{2}{\pi} V_{CC} \cdot I_{cm} - \frac{R'_L \cdot I_{cm}^2}{2} = P_{CC} - P_L$$

$I_{cm} = \dfrac{2}{\pi} \cdot \dfrac{V_{CC}}{R'_L}$ (The collector current at which the collector dissipation is maximum)

$$2 P_{cmax} = \frac{2}{\pi^2} \cdot \frac{V_{CC}^2}{R'_L}$$

Efficiency: $\quad \eta = \dfrac{P_L}{P_{CC}} = \dfrac{\pi}{4} \quad \dfrac{I_{cm}}{V_{CC}/R'_L}$

$$\boxed{\eta_{max} = \frac{\pi}{4} = 78.5\%}$$

Figure of merit: $\quad \dfrac{P_{cmax}}{P_{Lmax}} = \dfrac{2}{\pi^2} \approx \dfrac{1}{5}$

Power and efficiency variation:

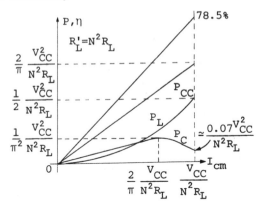

7.4 AMPLIFIERS USING COMPLEMENTARY SYMMETRY

The circuit:

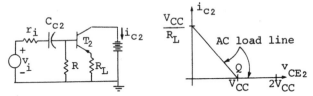

Circuit and load line for T_2 of complementary-symmetry emitter-follower

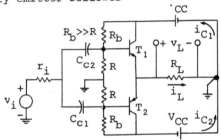

Complementary-symmetry amplifier

Each transistor is essentially a class "B" emitter follower.

$$i_L = i_{C_1} - i_{C_2}$$

$$I_{cm} = \frac{V_{CC}}{R_L} \quad \text{(If } v_i \text{ is sinusoidal, } i_L \text{ is also sinusoidal.)}$$

$$i_L = I_{CM} \cdot \sin \omega t$$

$$= \frac{V_{CC}}{R_L} \cdot \sin \omega t.$$

$$P_{L,max} = \frac{V_{CC}^2}{2R_L}$$

CHAPTER 8

FEEDBACK AMPLIFIERS

8.1 CLASSIFICATION OF AMPLIFIERS

8.1.1 VOLTAGE AMPLIFIER

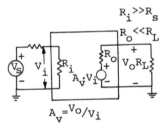

$$A_v = V_O/V_i$$

The voltage amplifier provides a voltage output proportional to the voltage input. The proportionality factor is independent of the source and load impedances. For the ideal amplifier:

$$R_i = \infty, \quad R_o = 0, \quad A_v = \frac{V_0}{V_s}$$

8.1.2 CURRENT AMPLIFIER

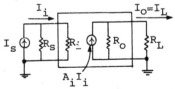

Norton's equivalent circuit of a current amplifier.

$$A_i = \frac{I_L}{I_i} \quad \text{(with } R_i = 0 \text{ representing the short-circuit current amplification.)}$$

$$R_i << R_s \quad \text{and} \quad R_o >> R_L$$

8.1.3 TRANSCONDUCTANCE AMPLIFIER

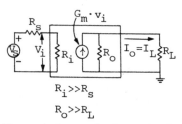

$$R_i >> R_s$$

$$R_o >> R_L$$

Thevenin's equivalent in its
input circuit and a Norton's
equivalent in its output circuit.

Output current is proportional to the signal voltage and is independent of R_S and R_L.

G_m (the short circuit mutual or transfer conductance)

$$= \frac{I_o}{v_i} \quad \text{for } R_L = 0$$

8.1.4 TRANSRESISTANCE AMPLIFIER

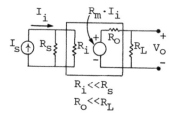

$$R_i << R_s$$
$$R_o << R_L$$

V_o is proportional to the signal current I_s and is independent of R_s and R_L.

$$R_m = \frac{V_o}{I_i} \text{ with } R_L = \infty \quad (R_m = \text{open circuit}$$
$$\text{mutual resistance})$$

8.1.5 IDEAL AMPLIFIER CHARACTERISTICS

	Amplifier type			
Parameter	Voltage	Current	Transconduct-ance	Transresist-ance
R_i	∞	0	∞	0
R_o	0	∞	∞	0

Transfer Charac-teristics $\quad V_o = A_v V_s \quad I_L = A_i I_s \quad I_L = G_m V_s \quad V_o = R_m I_s$

8.2 THE CONCEPT OF "FEEDBACK"

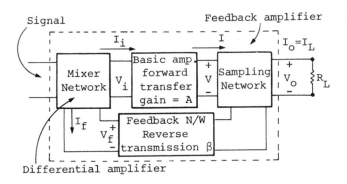

8.2.1 SAMPLING NETWORK

The output voltage is sampled by connecting the feedback network in shunt across the output voltage or node sampling.

Current or loop sampling - The feedback network is connected in series with the output.

8.2.2 TRANSFER RATIO OR GAIN

Each of the four quantities A_V, A_I, G_M and R_M is a different type of transfer gain of the basic amplifier without feedback (A).

A_f = The transfer gain with feedback. $= A_{vf}$ or A_{If} or G_{Mf} or R_{Mf}

$$A_{vf} = \frac{V_o}{V_s}, \quad \frac{I_o}{I_s} = A_{If}, \quad \frac{I_o}{V_s} = G_{Mf}, \quad \frac{V_o}{I_s} = R_{Mf}$$

8.3 TRANSFER GAIN WITH FEEDBACK

8.3.1 SINGLE-LOOP FEEDBACK AMPLIFIER

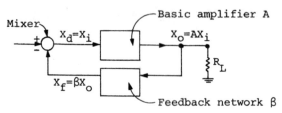

$$X_d = X_s - X_f = X_i = \text{The difference signal.}$$

8.3.2 FEEDBACK AMPLIFIER TOPOLOGIES

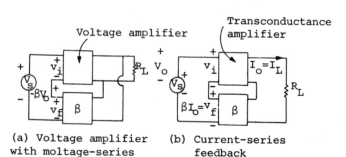

(a) Voltage amplifier with moltage-series feedback

(b) Current-series feedback

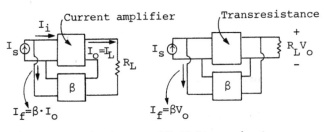

(c) Current-shunt
    feedback

(d) Voltage-shunt
    feedback

$\beta \equiv \dfrac{X_f}{X_o}$ β is a complex function of the signal frequency and is often a positive or negative real number.

$A = \dfrac{X_o}{X_i}$

$$A_f = \frac{X_o}{X_s} = \frac{A}{1 + \beta A}$$

If $|A_f| < |A|$, the feedback is termed negative; if $|A_f| > |A|$, the feedback is positive, or regenerative.

In the case of negative feedback, the gain of the ideal amplifier is divided by $|1 + \beta A|$, which exceeds unity.

Loop gain:

The loop gain $= -A \cdot \beta =$ The return ratio.

The return difference $D = 1 + A\beta$

$$N = \text{dB of feedback} = 20 \log \left| \frac{A_f}{A} \right| = 20 \log \left| \frac{1}{1 + A\beta} \right|$$

8.4 FEATURES OF THE NEGATIVE FEEDBACK AMPLIFIER

Sensitivity – The fractional change in amplification with feedback divided by the fractional change without feedback is the sensitivity of the transfer gain.

$$\frac{\left|dA_f/A_f\right|}{\left|\frac{dA}{A}\right|} = \frac{1}{\left|1 + \beta A\right|}$$

$$S = \text{Sensitivity} = \frac{1}{\left|1 + \beta A\right|}$$

Desensitivity = $D = 1 + \beta A$.

$A_f = A/D$

if $\left|\beta A\right| \gg 1$, $A_f = \dfrac{A}{1 + \beta A}$ $\dfrac{A}{\beta \cdot A} = \dfrac{1}{\beta}$

Frequency distortion – If the feedback network does not contain ractive elements, the overall gain is not a function of frequency.

Reduction of noise – The noise introduced in an amplifier is divided by the factor D if feedback is introduced.

8.5 INPUT RESISTANCE

Voltage–series feedback:

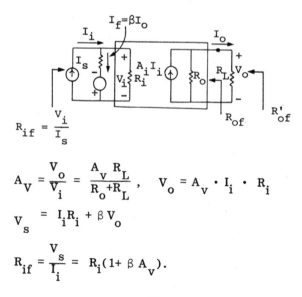

$R_{if} = \dfrac{V_i}{I_s}$

$A_V = \dfrac{V_o}{V_i} = \dfrac{A_V R_L}{R_o + R_L}$, $V_o = A_V \cdot I_i \cdot R_i$

$V_s = I_i R_i + \beta V_o$

$R_{if} = \dfrac{V_s}{I_i} = R_i (1 + \beta A_V)$.

$$A_v = \lim_{R_L \to \infty} A_v$$

Current-series feedback:

$$R_{if} = R_i(1 + \beta G_M), \quad G_M = \lim_{R_L \to 0} G_M$$

$$G_M = \frac{I_o}{V_i} = \frac{G_M \cdot R_o}{R_o + R_L}$$

Current-shunt feedback:

$$A_I = \frac{I_o}{I_i} = A_i \cdot R_o/(R_o + R_L)$$

$$I_s = (1 + \beta A_I)I_i$$

$$R_{if} = \frac{V_i}{(1 + \beta A_I)I_i} = \frac{R_i}{1 + \beta A_I}$$

$$A_i = \lim_{R_L \to 0} A_I$$

Voltage-shunt feedback:

$$R_{if} = \frac{R_i}{1 + \beta \cdot R_M}, \quad R_M = \frac{V_o}{V_i} = \frac{R_m \cdot R_L}{R_o + R_L}$$

$$R_m = \lim_{R_L \to \infty} R_M$$

8.6 OUTPUT RESISTANCE

For voltage sampling, $R_{of} < R_o$; whereas, for current sampling, $R_{of} > R_o$.

Voltage series feedback
$$R_{of} = \frac{V}{I} = \frac{R_o}{1 + \beta A_v}, \quad \text{where } D = 1 + \beta A_v$$

$$R_{of}' = R_o'/1 + \beta A_v \,, \quad \text{where } R_o' = R_o \parallel R_L.$$

Voltage-shunt feedback:

$$R_{of} = \frac{R_o}{1 + \beta R_m}, \quad R_{of}' = R_o'\Big/ 1 + \beta R_M$$

Current-shunt feedback:

$$R_{of} = \frac{V}{I} = R_o(1 + \beta A_i)$$

$$R_{of}' = R_o' \frac{1 + \beta A_i}{1 + \beta A_I}$$

$$= R_{of} \text{ (for } R_L = \infty, \ A_I = 0 \text{ and } R_o' = R_o).$$

Current-series feedback:
$$R_{of} = R_o(1 + \beta G_m) \text{ and } R_{of}' = R_o' \ \frac{1 + \beta G_m}{1 + \beta G_M}$$

8.7 FEEDBACK AMPLIFIER ANALYSIS

Topology Characteristic	Voltage shunt	Current shunt	Current series	Voltage series
Feedback signal X_f	Current	Current	Voltage	Voltage
Sample signal X_o	Voltage	Current	Current	Voltage
Input circuit: set†	$V_o = 0$	$I_o = 0$	$I_o = 0$	$V_o = 0$
Output circuit: set†	$V_i = 0$	$V_i = 0$	$I_i = 0$	$I_i = 0$
Signal source	Norton	Norton	Thévenin	Thévenin
$\beta = X_f/X_o$	I_f/V_o	I_f/I_o	V_f/I_o	V_f/V_o
$A = X_o/X_i$	$R_M = V_o/I_i$	$A_I = I_o/I_i$	$G_M = I_o/V_i$	$A_v = V_o/V_i$
$D = 1 + \beta A$	$1 + \beta R_M$	$1 + \beta A_I$	$1 + \beta G_M$	$1 + \beta A_v$
A_f	R_M/D	A_I/D	G_M/D	A_v/D
R_{if}	R_i/D	R_i/D	R_iD	R_iD
R_{of}	$\dfrac{R_o}{1 + \beta R_m}$	$R_o(1 + \beta A_i)$	$R_o(1 + \beta G_m)$	$\dfrac{R_o}{1 + \beta A_v}$
$R_{of}' = R_{of} \parallel R_L$	$\dfrac{R_o'}{D}$	$R_o' \dfrac{1 + \beta A_i}{D}$	$R_o' \dfrac{1 + \beta G_m}{D}$	$\dfrac{R_o'}{D}$

Analysis of a feedback amplifier:

Identify the topology – The input loop is defined as the mesh containing the applied signal voltage V_s and either (a) the base-to-emitter region of the first BJT, or (b) the gate-to-source region of the first FET, or (c) the section between the two inputs of a differential amplifier.

Mixing – There is a series mixing in the input circuit if there is a circuit compoment y in series with V_s and if y is connected to the output. If this is true, the voltage across y is the feedback signal $X_f = V_f$.

If this is not true, test for shunt comparison. Shunt mixing is present if the feedback signal subtracts from the applied excitation as a current at the input node.

Type of sampling:

A) Set V_o = 0 (i.e., set R_L = 0). If X_f becomes zero, the original system exhibited voltage sampling.

B) Set I_o = 0. If X_f becomes zero, current sampling was present in the original amplifier.

The amplifier without feedback – The amplifier configuration without providing feedback but with taking the loading of β network into account is obtained by applying following rules.

To find the input circuit:

A) Set V_o = 0 for voltage sampling. (i.e., short-circuit the output node)

B) Set I_o = 0 for current sampling. (i.e., open-circuit the output loop)

To find the output circuit :

A) Set V_i = 0 for shunt comparison.

B) Set I_i = 0 for series comparison. In other words, open-circuit the input loop

Outline of analysis – To find A_f, R_{if} and R_{of} the following steps are carried out:

A) Find out the topology first. It will determine whether X_f is a voltage or a current. The same applies to X_o .

B) Using the above rules, draw the basic amplifier circuit without feedback.

C) If X_f is a voltage, use a Thevenin's source; use Norton's source if X_f is a current.

D) Substitute the proper model for each active device (h-parameter model).

E) Indicate X_f and X_o on the circuit obtained by steps B, C, and D. Calculate $\beta = \dfrac{X_f}{X_o}$

F) Apply KVL and KCL to the equivalent circuit obtained after step D, to find A.

G) Calculate D, A_f, R_{if}, R_{of} and R'_{of} from A and B.

CHAPTER 9

FREQUENCY RESPONSE OF AMPLIFIERS

9.1 FREQUENCY DISTORTION

A plot of gain (phase) versus frequency of an amplifier is called the amplitude (phase) frequency-response characteristic.

If the amplification A is independent of frequency, and if the phase shift θ is proportional to frequency (or is zero), then the amplifier will preserve the form of the input signal, although the signal will be shifted in time by an amount θ/ω.

Frequency-response characteristics:

A) Low-frequency region: The amplifier behaves like a simple high-pass circuit.

B) Mid-band frequency: Amplification and delay is quite constant. The gain is normalized to unity.

C) High-frequency region: The circuit behaves like a low-pass network.

Low and high-frequency response:

Low-frequency response:

The low-frequency region is like a simple high-pass circuit.

$$V_o = \frac{R_1}{R_1 + \frac{1}{SC_1}}, \quad V_i = \frac{S}{S + \frac{1}{R_1 C_1}} V_i$$

$$|A_L(f)| = \frac{1}{1-j(f_L/f)}, \quad f_L = \frac{1}{2\pi R_1 C_1}, \quad \theta_L = \text{arc tan} \frac{f_L}{f}$$

The low-3dB-frequency: At this frequency, the gain has fallen to 0.707 times its mid-band value A_o; $f = f_L$ is the low-3dB frequency.

High-frequency response:

$$V_o = \frac{\frac{1}{SC_2}}{R_2 + \frac{1}{SC_2}} \quad V_i = \frac{1}{1 + SR_2 C_2} V_i$$

$$|A_H(f)| = 1/\sqrt{1+(f/f_H)^2} \quad \theta_H = -\text{arc tan} \frac{f}{f_H}$$

$$f_H = 1/2\pi R_2 C_2$$

The high-3dB frequency: At $f = f_H$, the gain is reduced to $1/\sqrt{2}$ times its mid-band value, this is called the high-3dB frequency.

Bandwidth: B.W. = frequency range from f_L to f_H.

9.2 THE EFFECT OF COUPLING AND EMITTER BYPASS CAPACITORS ON LOW-FREQUENCY RESPONSE

An emitter resistance R_e is used for self-bias in an

amplifier and to avoid degeneration; C_Z is used to bypass R'_e.

$$V_o = -I_b h_{fe} R_c = \frac{-V_s \cdot h_{fe} \cdot R_c}{R_s + h_{ie} + Z_b + Z'_e}$$

$$Z_b = \frac{1}{j\omega c_b} \quad \text{and} \quad Z'_e = (1+h_{fe}) \frac{R_e}{1+j\omega C_z \cdot R_e}$$

$$j\omega(Z_b + Z'_e) = \frac{1}{C_b} + \frac{1+h_{fe}}{C_Z} = \frac{1}{C_1}$$

$$C_1 = C_b + \frac{C_Z}{1+h_{fe}}$$

$$A_{vs} = \frac{V_o}{V_s} = \frac{-h_{fe} R_c}{R_s + h_{ie}} \cdot \frac{1}{1 - j/\omega c_1 (R_s + h_{ie})}$$

$$A_o = \text{mid-band gain} = \frac{-h_{fe} \cdot R_c}{R_s + h_{ie}}, \quad \frac{A_{vs}}{A_o} = \frac{1}{1 - j(f_L/f)}$$

Low-3dB freq. $f_L = 1/2\pi C_1 (R_s + h_{ie})$ (at f_L, $R_s + h_{ie} = R_1$)

83

9.3 THE HYBRID-π TRANSISTOR MODEL AT HIGH FREQUENCIES

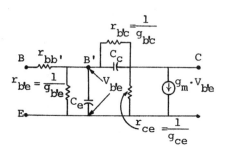

$C_c = C_{b'c}$ is the measured C_B ouput capacitance with the input open ($I_E = 0$). (C_c is the transition capacitance, which varies as V_{CB}^{-n}, where n is $\frac{1}{2}$ or $\frac{1}{3}$). $C_e = C_{De} + C_{Te}$, where C_{De} is the emitter diffusion capacitance and C_{Te} is emitter-junction capacitance.

The diffusion capacitance:

$$C_{De} = g_m \ \frac{w^2}{2D_B} \quad (D_B = \text{the diffusion constant for minority carriers})$$

Simplified model:

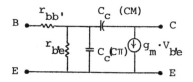

Variations of Hybrid-π barameters:

Dependence of parameters upon current, voltage and temperature

Variation with increasing:

Parameter	$\lvert I_c \rvert$	$\lvert V_{CE} \rvert$	T
g_m	$\lvert I_c \rvert$	Independent	$1/T$
$r_{bb'}$	Decreases		Increases
$r_{b'e}$	$1/\lvert I_c \rvert$	Increases	Increases
C_e	$\lvert I_c \rvert$	Decreases	
C_c	Independent	Decreases	Independent
h_{fe}		Increases	Increases
h_{ie}	$1/\lvert I_c \rvert$	Increases	Increases

9.4 THE C-E SHORT-CIRCUIT CURRENT GAIN

Approximate equivalent circuit:

$$I_L = -g_m \ V_{b'e}$$

$$V_{b'e} = I_i/g_{b'e} + j\,\omega(C_e + C_c)$$

$$A_i = \frac{I_L}{I_i} = \frac{-g_m}{(g_{b'e} + j\,\omega[C_e + C_c])} = \frac{-h_{fe}}{1 + j(f/f_\beta)}$$

$$f_\beta = \frac{g_{b'e}}{2\pi(C_e + C_c)} = \frac{1}{h_{fe}} \frac{g_m}{2\pi(C_e + C_c)}$$

At $f = f_\beta$, $|A_i|$ is equal to $1/\sqrt{2} = 0.707$ of its low-frequency value, h_{fe}.

B.W. = The frequency range up to f_β.

The parameter f_T:

f_T is the frequency at which the short-circuit common-emitter current gain attains unit magnitude.

$$f_T \approx h_{fe} \cdot f_\beta = \frac{g_m}{2\pi(C_e + C_c)} \approx \frac{g_m}{2\pi \cdot C_e}$$

Variation of f_T on collector current:

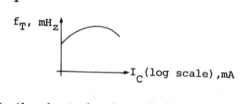

f_T also represents the short-circuit current gain-bandwidth product.

The plot of the short-circuit CE current gain versus the frequency:

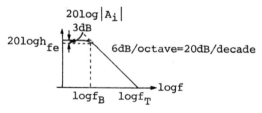

At f_T (the highest frequency of interest),

$$\frac{\omega C_c}{g_m} = \frac{2\pi \cdot f_T \, C_c}{g_m} = \frac{C_c}{C_e + C_c} \cong 0.03$$

9.5 THE COMMON-DRAIN AMPLIFIER AT HIGH FREQUENCIES

The source-follower and its small-signal high-frequency equivalent circuit:

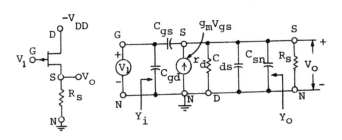

$$A_v = \text{The voltage gain} = \frac{(g_m + j\omega C_{gs})R_s}{1 + (g_m + g_d + j\omega C_T)R_s}$$

$$C_T = C_{gs} + C_{ds} + C_{sn}$$

$$A_v (\text{at low frequency}) \approx \frac{g_m \cdot R_s}{1 + (g_m + g_d)R_s} \approx \frac{g_m}{g_m + g_d}$$

$$= \frac{\mu}{1 + \mu}$$

Input and output admittance:

$$y_i = j\omega C_{gd} + j\omega C_{gs}(1 - A_v) \approx j\omega C_{gd}$$

$$y_o = g_m + g_d + j\omega C_T$$

$$R_o (\text{at low frequency}) = 1/g_m + g_d \approx 1/g_m$$

9.6 THE COMMON-SOURCE AMPLIFIER AT HIGH FREQUENCIES

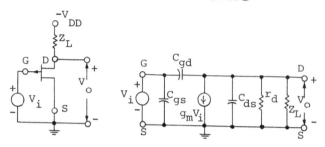

$$I = -g_m \cdot V_i + V_i \cdot y_{gd}$$

The amplification is $\displaystyle A_v = \frac{V_o}{V_i} = \frac{I \cdot Z}{V_i} = \frac{-g_m + y_{gd}}{y_L + g_d + y_{ds} + y_{gd}}$

$$A_v \text{ (at low frequencies)} = \frac{-g_m}{y_L + g_d} = \frac{-g_m \cdot r_d \cdot Z_L}{r_d + Z_L} = -g_m \cdot Z_L$$

The input admittance is

$$y_i = y_{gs} + (1-A_v)y_{gd} = G_i + j\omega C_i \approx j\omega C_{gd}$$

9.7 THE EMITTER-FOLLOWER AMPLIFIER AT HIGH FREQUENCIES

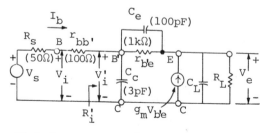

High-frequency equivalent circuit

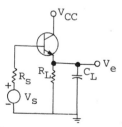

Emitter-follower

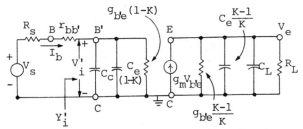

Fig: Equivalent circuit using Miller's theorem

The nodal equations at the nodes of B' and E are:

$$G'_s V_s = [G'_s + g_{b'e} + s(C_c + C_e)]v'_i - (g_{b'e} + sC_e)V_e$$

$$0 = -(g + sC_e)V'_i + [g + \frac{1}{R_L} + s(C_e + C_L)]V_e$$

$$G'_s = 1/R_s + r_{bb'} \quad \text{and} \quad g = g_m + g_{b'e}$$

Single pole solution:

$$K = V_e / V'_i$$

$$V_e = \frac{g_m \cdot V_{b'e}}{\frac{1}{R_L} + j\omega C_L} = \frac{g_m \cdot R_L (V'_i - V_e)}{1 + j\omega C_L R_L} \quad \text{(for K=1)}$$

$$K = \frac{K_o}{1 + jf/f_H}, \quad K_o = \frac{g_m \cdot R_L}{1 + g_m \cdot R_L} \approx 1.$$

$$f_H = \frac{1 + g_m \cdot R_L}{2\pi \cdot C_L \cdot R_L} \approx \frac{g_m}{2\pi C_L} = \frac{f_T \cdot C_e}{C_L}$$

Input admittance:

$$Y'_i = \frac{I_b}{V'_i} = j\omega [C_c + (1-K)C_e] + (1-K)g_{b'e}$$

$$Y'_i = j2\pi f \cdot C_c + (g_{b'e} + j2\pi \cdot f \cdot C_e) \frac{1 - K_o + jf/f_H}{1 + jf/f_H}$$

$$Y'_i = j2\pi f \cdot C_c + j\, g_{b'e} \cdot f/f_H - 2\pi \cdot f^2 \frac{C_e}{f_H}$$

9.8 SINGLE-STAGE CE TRANSISTOR AMPLIFIER RESPONSE

Equivalent circuit:

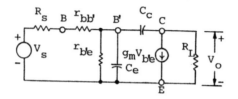

The transfer function:

$$\frac{V_o}{V_s} = \frac{-G'_s \cdot R_L (g_m - {}^sC_c)}{s^2 C_e \cdot C_c \cdot R_L + {}^s [C_e + C_c + C_c R_L (g_m + g_{b'e} + G'_s)] + G'_s}$$

The transfer function is of the form

$$A_{vs} = \frac{V_o}{V_s} = \frac{K_1(s-s_0)}{(s-s_1)(s-s_2)}$$

Approximate analysis:

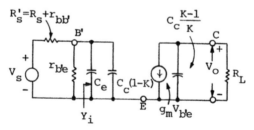

After application of Miller's theorem

$K = \frac{V_{ce}}{V_{b'e}}$, for $|K| >> 1$. Also the output capacitance is C_c and the output time constant is $C_c \cdot R_L$.

$$K = -g_m \cdot R_L \text{ (neglecting } C_c)$$

Input capacitance $C = C_e + C_c(1 + g_m \cdot R_L)$

Input loop resistance $= R = R'_s \parallel r_{b'e}$

$$A_{vs} = v_o/v_s = (-g_m \cdot R_L \cdot G'_s)/(G'_s + g_{b'e} + sC)$$

$$A_{vs} = \frac{A_{vso}}{1 + j\, f/f_H}$$

The high 3-dB frequency $= f_H = \dfrac{G'_s + g_{b'e}}{2\pi C} = \dfrac{1}{2\pi R \cdot C}$

$$\left| A_{vs} \right| = \frac{\left| A_{vso} \right|}{[1 + (f/f_H)^2]^{\frac{1}{2}}} \;,\quad \theta_1 = -\pi - \arctan f/f_H$$

HANDBOOK OF
MATHEMATICAL,
SCIENTIFIC, and
ENGINEERING
FORMULAS, TABLES,
FUNCTIONS, GRAPHS,
TRANSFORMS

A particularly useful reference for those in math, science, engineering and other technical fields. Includes the most-often used formulas, tables, transforms, functions, and graphs which are needed as tools in solving problems. The entire field of special functions is also covered. A large amount of scientific data which is often of interest to scientists and engineers has been included.

Available at your local bookstore or order directly from us by sending in coupon below.

THE PROBLEM SOLVERS

The "PROBLEM SOLVERS" are comprehensive supplemental textbooks designed to save time in finding solutions to problems. Each "PROBLEM SOLVER" is the first of its kind ever produced in its field. It is the product of a massive effort to illustrate almost any imaginable problem in exceptional depth, detail, and clarity. Each problem is worked out in detail with step-by-step solution, and the problems are arranged in order of complexity from elementary to advanced. Each book is fully indexed for locating problems rapidly.

ADVANCED CALCULUS
ALGEBRA & TRIGONOMETRY
AUTOMATIC CONTROL
  SYSTEMS/ROBOTICS
BIOLOGY
BUSINESS, MANAGEMENT,
  & FINANCE
CALCULUS
CHEMISTRY
COMPLEX VARIABLES
COMPUTER SCIENCE
DIFFERENTIAL EQUATIONS
ECONOMICS
ELECTRICAL MACHINES
ELECTRIC CIRCUITS
ELECTROMAGNETICS
ELECTRONIC COMMUNICATIONS
ELECTRONICS
FINITE & DISCRETE MATH
FLUID MECHANICS/DYNAMICS
GENETICS

GEOMETRY:
PLANE · SOLID · ANALYTIC
HEAT TRANSFER
LINEAR ALGEBRA
MACHINE DESIGN
MECHANICS : STATICS · DYNAMICS
NUMERICAL ANALYSIS
OPERATIONS RESEARCH
OPTICS
ORGANIC CHEMISTRY
PHYSICAL CHEMISTRY
PHYSICS
PRE-CALCULUS
PSYCHOLOGY
STATISTICS
STRENGTH OF MATERIALS &
  MECHANICS OF SOLIDS
TECHNICAL DESIGN GRAPHICS
THERMODYNAMICS
TRANSPORT PHENOMENA :
MOMENTUM · ENERGY · MASS
VECTOR ANALYSIS

If you would like more information about any of these books, complete the coupon below and return it to us or go to your local bookstore.

RESEARCH and EDUCATION ASSOCIATION
61 Ethel Road W. • Piscataway • New Jersey 08854
Phone: (201) 819-8880

Please send me more information about your Problem Solver Books

Name _____

Address _____

City _____ State _____ Zip _____